M2A3

Bradley Fighting Vehicle, Vol. 2

In Detail

A photo-reference journal of the United States Army's Bradley Fighting Vehicle based in CONUS (Continental United States). Includes more than 140 pages, color profiles and several hundred photographs.

SABOT Publications

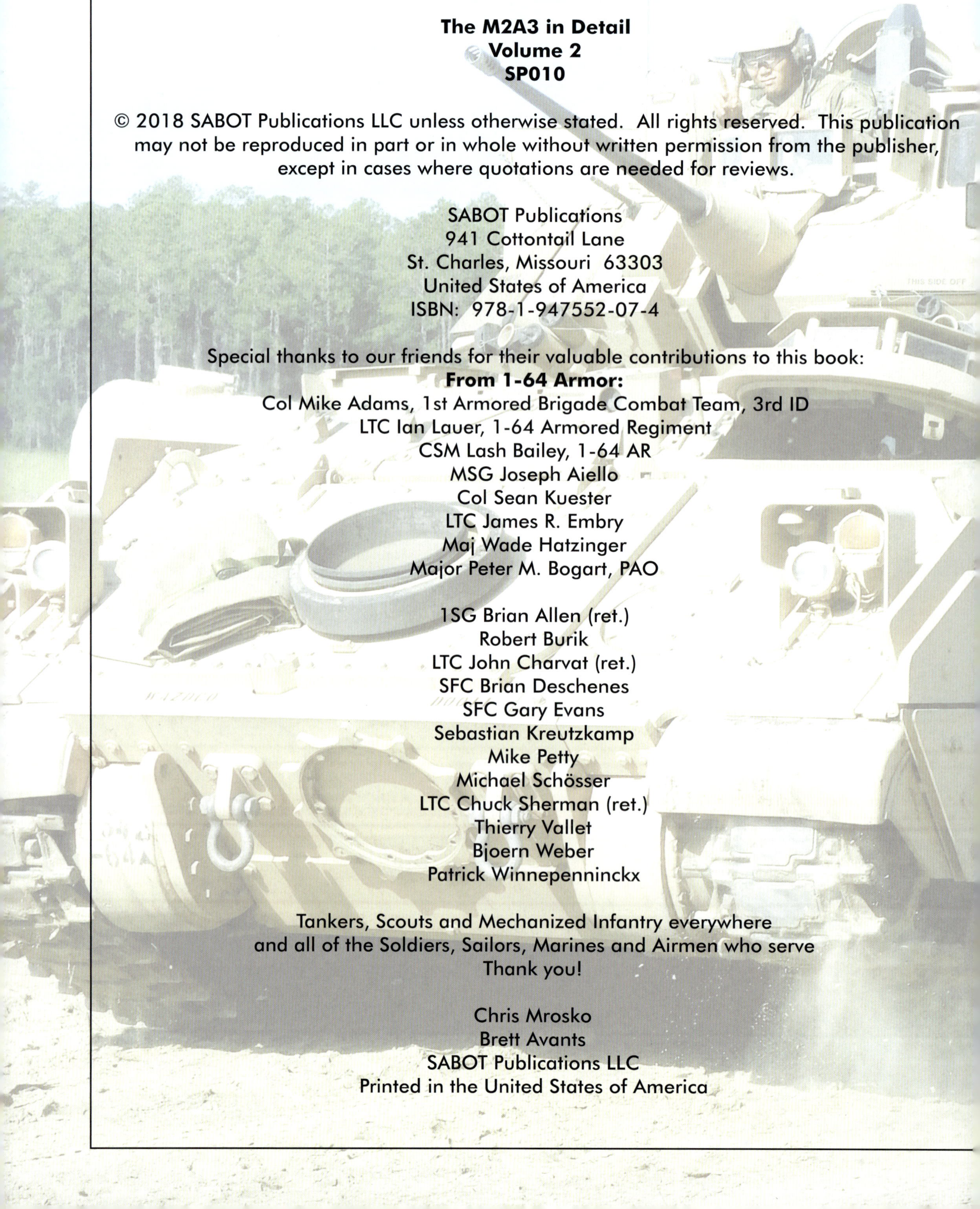

The M2A3 in Detail
Volume 2
SP010

SABOT Publications
941 Cottontail Lane
St. Charles, Missouri 63303
United States of America
ISBN: 978-1-947552-07-4

Special thanks to our friends for their valuable contributions to this book:
From 1-64 Armor:
Col Mike Adams, 1st Armored Brigade Combat Team, 3rd ID
LTC Ian Lauer, 1-64 Armored Regiment
CSM Lash Bailey, 1-64 AR
MSG Joseph Aiello
Col Sean Kuester
LTC James R. Embry
Maj Wade Hatzinger
Major Peter M. Bogart, PAO

1SG Brian Allen (ret.)
Robert Burik
LTC John Charvat (ret.)
SFC Brian Deschenes
SFC Gary Evans
Sebastian Kreutzkamp
Mike Petty
Michael Schösser
LTC Chuck Sherman (ret.)
Thierry Vallet
Bjoern Weber
Patrick Winnepenninckx

Tankers, Scouts and Mechanized Infantry everywhere
and all of the Soldiers, Sailors, Marines and Airmen who serve
Thank you!

Chris Mrosko
Brett Avants
SABOT Publications LLC
Printed in the United States of America

Introduction

The M2/M3 A3 Bradley Fighting Vehicle (BFV) represents the newest version of the fighting vehicle, and the latest in a series of Bradleys. The A3 Bradley, like its predecessors, has both Infantry (M2 A3) (IFV) and Cavalry (M3 A3) (CFV) models. The primary differences between them are related to their missions.

The first Bradley, fielded in late 1982 and generally called the AO or the vanilla Bradley, represented a huge change from the M113 armored personnel carrier. The BFV, like the M113, was designed to carry personnel, but differed markedly from the M113 in its ability to defeat armored and unarmored vehicles with its 25 mm Bushmaster cannon, and defeat personnel targets and unarmored targets using the 7.62 coaxially mounted machinegun. It also became a tank killer with the Tube-launched Optically-tracked, Wire-guided (TOW) missile. The original Bradley was also noteworthy for the difficulties encountered in the effort to get it accepted. As each version has been introduced, some have derided the vehicle, maintaining that it is not needed, or simply finding fault with it. The Bradley has lasted, however. The primary shortfall at this point may be a failure to learn from the lessons from the past.
The original BFV was improved, in the 1986 A1 variant: with the TOW II, the gas particulate filter unit, etc. The next version was the 1988 A2 that went to battle in Operation Desert Storm (ODS). Most equipment and performance deficiencies identified or confirmed during Desert Storm were subsequently remedied in the improved A2, the M2A2 ODS variant, more commonly referred to as the ODS. The ODS improvements brought the Bradley more in line with the Abrams Main Battle Tank, and addressed some of the deficiencies which had been identified in the Bradley from the very beginning. The success of the Bradley during Desert Storm brought the vehicle renewed prominence, and public praise for its performance. Barely 10 years after the war, a new Bradley was introduced. Unlike the previous versions, the M2/M3 A3 vehicle is radically different from those that preceded it.

This book is a visual reference on the M2A3 Bradley, focusing on Bradleys in the Continental United States (CONUS) as well as a few in Europe and Korea. Color profiles include a unique set of concept art for an Operational Camouflage Pattern (OCP) camouflaged Bradley for the 199th Infantry Brigade at Ft. Benning, Georgia.

The M2A3

An M2A3 Bradley from Company B, 1-64 Armor Regiment "Desert Rogues," 3rd Infantry Division, fires on its first engagement during gunnery Table VI.

BELOW: Soldiers of Company B, 1-64 Armor "Desert Rogues," 3rd Infantry Division, maneuver their M2A3 Bradley during an exportable combat training center (XCTC) training event located at Camp Shelby, Mississippi. At Camp Shelby, the Rogues act as the opposing force in support of the 278th Armored Cavalry Regiment, a National Guard unit headquartered in Knoxville, Tennessee.

ABOVE & BELOW: An M2A3 Bradley from Company A, 2-8 Cavalry Regiment, 1st Cavalry Division, begins a qualification table at Adazi Training Area, Latvia, November 11, 2014.

ABOVE: A Soldier in the Oregon National Guard's 3rd Combined Arms Battalion, 116th Cavalry Regiment, sits in the driver's seat of an M2A3 Bradley. BELOW: Two M2A3 Bradleys from Company A, 2-8 Cavalry Regiment, 1st Cavalry Division, prepare to start a live-fire training exercise at Adazi Training Area, Latvia, on November 21, 2014.

ABOVE: A M2A3 Bradley from Company B, 2-8 Cavalry Regiment, 1st Cavalry Division, maneuvering with Latvian troops at Adazi Training Area, Latvia.
RIGHT: Members of Troop C, 1-98th Cavalry Regiment, 155th Armored Brigade Combat Team, Mississippi Army National Guard prepare for training during an Exportable Combat Training Capability exercise at Camp Shelby Mississippi.

ABOVE: From a hasty fighting position, U.S. Soldiers fire a M240C coaxial machine gun from a Bradley to repel an advance of British forces during a situational training exercise. The training is part of the multinational exercise Iron Sword 2014 involving 2,500 Soldiers from Canada, the Czech Republic, Estonia, Germany, Hungary, Lithuania, Luxembourg, the United States and the United Kingdom. BELOW: A M2A3 Bradley with the 10th Engineer Battalion simulates firing a TOW missile during their Gunnery Table XII at Fort Stewart, Georgia December, 2016.

C21 of 1-37 Armor moves to a firing position during Bradley gunnery.

BELOW: Aerial drone image of an M2A3 Bradley from the 1st Armor Brigade Combat Team, 3rd Infantry Division, conducting Table XII gunnery at Fort Stewart, Georgia. December, 2016.

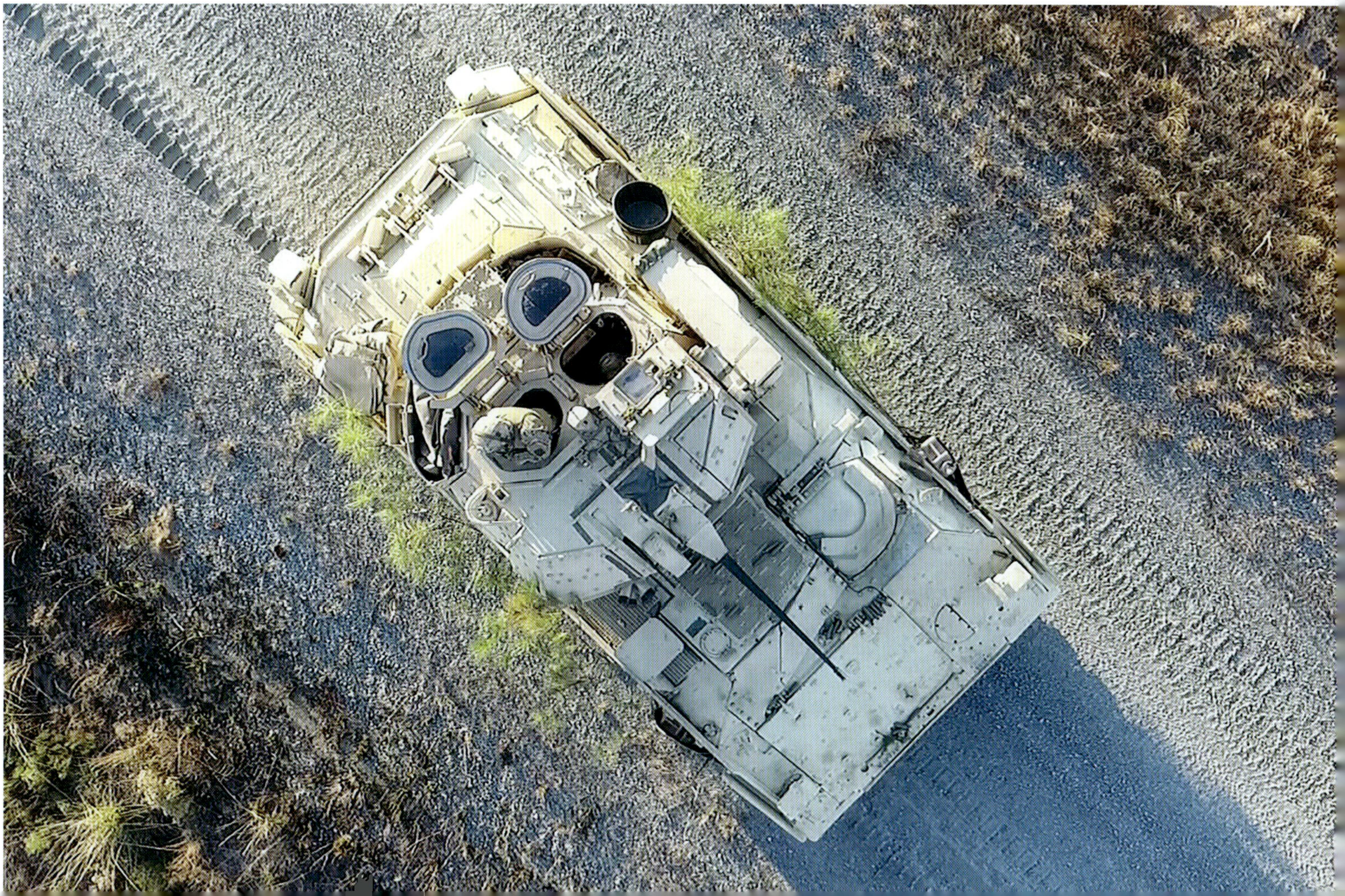

An M2A3 Bradley with Company D, 1st Battalion, 155th Infantry Regiment moves into position to support an assault while participating in an Exportable Combat Training Capability exercise at Camp Shelby Joint Forces Training Center, near Hattiesburg, Mississippi.

8th Cavalry Regiment soldiers dismount a Bradley and advance on a target as Hungarian Solders in BTR-80 Armored Personnel Carriers and dismounted Hungarian Soldiers provide suppressing fire.

A M2A3 Bradley waits to cross over the Kwisa River during a multi-national gap crossing exercise held in Poland on November 16, 2017.

A Bradley with 1-8 Infantry, 4th Infantry Division during life fire exercises.

An M2A3 belonging to 3rd Battalion, 29th Field Artillery Regiment, 3rd Armored Brigade Combat Team, 4th Infantry Division prepares to conduct a multinational combined arms breach of obstacles training exercise Feb. 15, at the Gora Hetmanska Range in Drawsko Pomorskie.

M2A3 Bradley during Decisive Action Rotation 17-05 at the National Training Center in Fort Irwin, California, April 5, 2017.

A cartridge ejects from a Bradley during a Live Fire Accuracy Screening Test in Swietozow, Poland, January 16, 2017.

U.S. Soldiers assigned to 1st Battalion, 8th Infantry Regiment, 3rd Armored Brigade Combat Team, 4th Infantry Division, leave the qualification area with their Bradley at a live fire range at Karliki range, Zagan, Poland, February 2, 2017.

An M2A3 Bradley Fighting Vehicle crew with Chaos Company, 1st Battalion, 68th Armor Regiment, 3rd Armored Brigade Combat Team, 4th Infantry Division, moves to engage a target during a platoon live-fire exercise at the Grafenwoehr Training Area, Germany, July 28, 2017.

In order to break up the desert camouflage during a cooperative training exercise in Sweden September of 2017, American vehicle crews of the 1st Combined Arms Battalion – 194th Armor adapted an old technique by borrowing some foliage from the Swedish countryside.

An M2A3 with Chaos Company, 1st Battalion, 68th Armor Regiment, 3rd Armored Brigade Combat Team, 4th Infantry Division, following the completion of a platoon live-fire exercise at Grafenwoehr Training Area, Germany, July 28, 2017.

U.S. Soldiers with Bravo Company, 1st Battalion, 8th Infantry Regiment, 3rd Armored Brigade Combat Team, 4th Infantry Division, conduct a combined arms live-fire exercise at Grafenwoehr Training Area, Germany, August 18, 2017.

M2A3 Walkaround

1CAV
2-5CAV
B-14
740

Front

This 10th Engineer, 3rd ID M2A3 Bradley sits in the bullpen at Ft Stewart waiting for a turn during a Combined Arms Field Exercise (CALFEX). The 600HP Cummins VTA-903T diesel engine undergoes some maintenance during the downtime.

ABOVE: PROMETHEUS is the brigade commander's track. He generally has a Bradley as his combat vehicle in a combined arms brigade, known as a Brigade Combat Team (BCT).

RIGHT: Another 10th Engineer M2A3 Bradley parked in the bullpen. The mudflaps, which are rubber, are usually painted over in the CARC tan paint, but over time the paint chips off the rubber.

ABOVE: This M2A3 sits near a fuel point waiting for the order to commence a road march to attack a prepared position during a CALFEX. Note the baggage on the side of the vehicle, and the extra roadwheel on the front slope. An extra sprocket is also attached to the lower front hull. LEFT: The driver's hatch in the open-protected mode. This close-up of the driver's hatch shows the anti-laser coating to good effect. Depending on the view angle, it can appear purple, red, blue, or gold. The driver can see out but the hatch still serves to protect the driver somewhat. The front headlight assembly is also prominent in this photo.

ABOVE: A head-on view of the M2A3 Bradley. Again, this vehicle has a spare roadwheel on the front slope and a spare sprocket between the tow shackles. These vehicles also have the T-157i track. RIGHT: The driver has hung his Combat Vehicle Crewman (CVC) helmet from one of the two anti-foul poles on either side of his hatch. It prevents the wires from the TOW missiles from fouling around the driver's hatch. The inside of the driver's hatch is also visible in this photo. The driver's CVC Helmet is well worn.

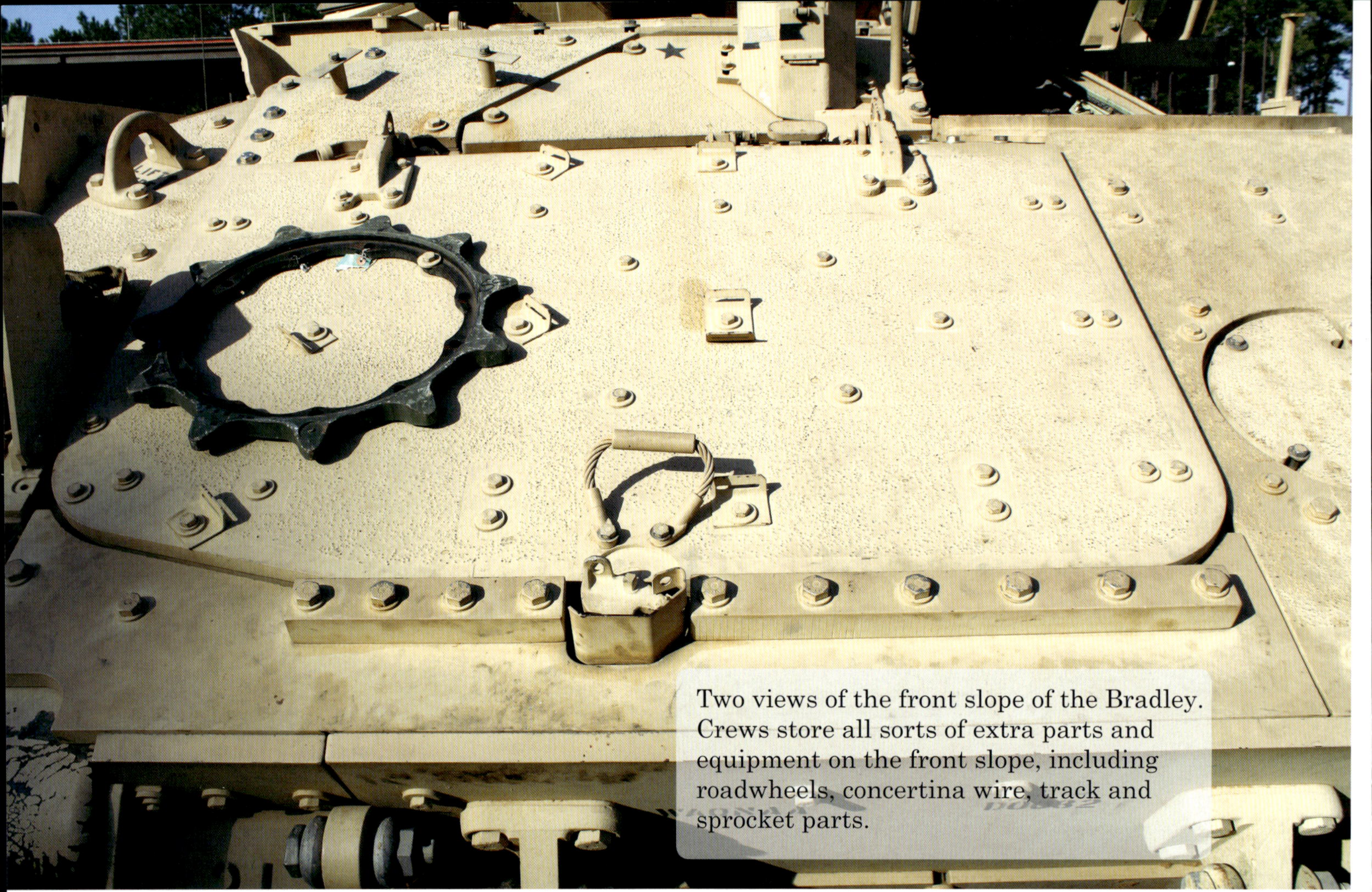

Two views of the front slope of the Bradley. Crews store all sorts of extra parts and equipment on the front slope, including roadwheels, concertina wire, track and sprocket parts.

At the NTC crews usually carry coils of razor wire in order to be ready to set up snap checkpoints. This view also shows the Driver's Vision Enhancer (DVE) to the right of the driver's hatch.

The black box behind the driver's hatch is part of the Multiple Integrated Laser Engagement System (MILES), which is used during wargame exercises.

Detailed view of the interior of the driver's hatch.

ABOVE: Another view of the front slope on the M2A3 Bradley.

LEFT: The front left sprocket. Also note the mud flap was originally painted tan with the rest of the vehicle, but the paint is quickly wearing off the rubber mud flap. Also notice how little chipping there is on the rest of the vehicle. The U.S. Army CARC Tan is pretty durable, even in desert conditions. The M2A3 uses a different front mudflap than its predecessors.

ABOVE: A close-up of the left front headlight assembly of the M2A3.

RIGHT: Another close view of the left drive sprocket. The tracks on this vehicle are well-worn. The metal has rusted, and the track pads have worn down close to the metal tracks.

Right Side

ABOVE and LEFT: The exhaust of the M2A3. This area of the vehicle is well-protected and diverts the exhaust away from the commander's station. In Desert Storm, due to the wind and to the early Bradley design, the exhaust would constantly be in the commander's face.

ABOVE: An overall view of the MILES set up from the right side. Note that the system has been upgraded over the years and so different units sometimes show different MILES gear. In addition to the laser emitter over the main gun, the receiver nodules can be seen on the turret side, and the "Whoopie Light" on the rear turret to indicate a knocked out vehicle.

RIGHT: Both sides of the M2A3 have this rubber brush guard to store and protect gear.

ABOVE and LEFT: Two nice views of the crew's packs being stowed on the outside of the vehicle. A3 Bradleys have hooks and straps to secure equipment on the outside of the vehicle. On earlier Bradleys, the crews would unscrew the bolts for the add-on armor, and literally screw the pack's strap into the hole with the bolt. Interior space and bustle rack is limited, so as much of the baggage as possible is stored on the outside. The rear mudflaps commonly get fouled on the tracks and bend like here.

Lots of dirt build up from pivoting on soft soil. You can see where the mud has splattered up on the side skirts and the rear storage box. Note the dirt and mud in the roadwheels also. BELOW: A view of how gear is hung from the side of the vehicle.

ABOVE: Red dirt and clay on the roadwheels of this Bradley. The side skirts look pretty clean though!

LEFT: The Combat Identification Panel (CIP) on this Bradley. These panels are designed to produce a distinct and easily identifiable infrared signature when seen through thermal imaging systems. Originally created as a hasty expediency, they actually work very well. The physical separation from the body of the vehicle meant that the panel would appear as a contrasting dark (i.e. cooler) area through thermal viewers. This CIP Panel has the vehicle number applied in black tape.

ABOVE: This M2A3 has been reassigned to B Company as seen on the Combat Identification Panel (CIP) on the side of the vehicle. BELOW: This M2A3 carries the spare sprocket on the lower front hull. A drip pan is shoved behind the light cluster.

Left Side

ABOVE: The orange cones indicate this M2A3 is an engineer Bradley.
BELOW: The Bradley Commander (BC) dismounts from the vehicle in order to ground guide the driver into the bullpen. The main gun always faces downrange as a safety measure. This vehicle does not have the CIP mounted, but the frame is evident.

ABOVE: The dismounts wait for the order to start the road march.

RIGHT: An M2A3 at NTC ready to move out. Note the abundance of orange cones in the bustle rack of the turret.

This B Company track does not have the small rear CIPs mounted. The empty frames are seen on the sponson boxes.

31
10E
WAP7B0
D0011
B11

Rear

ABOVE: The rear of the Bradley with the ramp down. RIGHT: The right rear sponson box is filled with light sticks and wire spools. This is probably an engineer track since only they usually carry this many glow sticks.

ABOVE: The rear of the Bradley showing the ramp open. The straps that hold the packs are visible on the sides. RIGHT: The Brigade Commander's track right rear sponson includes an oil can.

ABOVE: While waiting in the bullpen, the brigade commander's crew breaks down the 25MM chain gun. A block of wood is commonly used to prop up the rear ramp to keep it level. PAGE 43: Detail of the left rear of the vehicle. The clasps for securing the sponson boxes closed are rubber, but painted tan here.

The M242 Bushmaster is a 25 mm (25×137mm) chain-driven autocannon. It is used extensively by the U.S. military, as well as by NATO's and some other nations' forces in ground combat vehicles, such as the M2/M3 series and various watercraft.

It is an externally powered, chain-driven, single-barrel weapon which may be fired in semi-automatic, burst, or automatic modes. It is fed by a metallic link belt and has dual-feed capability. The term "chain gun" derives from the use of a roller chain that drives the bolt back and forth. The gun can fire both armor piercing (AP) rounds as well as high explosive (HE) rounds. The AP rounds can destroy lightly armored targets, while the HE rounds are like small 25mm grernades that can supress and destroy enemy infantry and soft targets. In Operation Desert Storm, one of the authors of this book used his 25mm gun firing AP ammunition to destroy several T-55 series tanks at close range (too close, actually). The AP rounds penetrated the turret armor and caused brew ups on several occasions. With over 10,000 units sold worldwide, it is one of the most successful modern autocannons.

3 1-64

WAZDCO
D0024

PAGE 44 ABOVE: An M2A3 stops at a Battle Position (BP) to take on range targets, commonly called the Plywoodian Army, as the targets are made of plywood. The TOW system is only raised when the vehicle is in a stopped position. PAGE 44 BELOW: One of the distinguishing aspects of the M2A3 vs the M3A3 is the location of the periscopes on the upper rear hull of the vehicle. On the M2A3 they are on the hull itself. On the M3A3 they are on the troop hatch. PAGE 45 ABOVE: This battalion S-3 vehicle is in a hide position during a CALFEX. The HQ element is typically stationary during a CALFEX but all vehicles and soldiers are "Tactical." RIGHT: The inside view of the S-3 track showing the blast resistant seats in the M2A3.

FIRE
EXTINGUISHER

ABOVE PAGE 46 & 47: Two close-up views of the rear of the Bradley. Note the rear fire extinguisher on Page 46, and the various tie downs and brackets.

PAGE 46 BELOW: Bradleys begin to line up behind the tanks for the road march to the attack position.

PAGE 47 BELOW: Another view of packs strapped to the side of the Bradley.

The receiver nodules of the MILES system are different on the top photo of the Bradley vs the one on the bottom photo. The top photo is the newest incarnation of the system.

Turret

Looking up at the commander's Independent Viewer (CIV).

ABOVE: In the US Army tradition dictates that vehicle names begin with the company letter. So this vehicle is an A Company vehicle. The name only appears on the TOW box.

LEFT: A different kind of CIP is on the Bradley turret in this photo, and it is shown here in its folded position.

ABOVE: Spent shells from the M242 25mm chain gun are ejected out of the bottom right side of the turret. In this angle the folded CIP is seen as well.

BELOW: The rear of the turret.

ABOVE: Close-up of a Velcro attachment for a part of the MILES gear. BELOW: A shot of the front of the turret, showing the sensors for the MILES gear and the gun elevated.

The CIP on this example is folded open, and shows some wear.

PAGES 54-55: In these views of the turret facing backwards, the rear of the turret shows the new MILES receiver nodules. The older MILES Velcro belt glue marks are still evident on the lower portion of the turret.

This different Bradley shows similar MILES evidence, but in a slightly different area of the turret. The black Velcro is seen on the vehicle for the new receiver nodules.

ABOVE and BELOW: A good close-up of the older MILES laser emitter and the belts with the receiver nodules.

MILES: The multiple integrated laser engagement system, or MILES, is used by the U.S. military and other armed forces around the world for training purposes. It uses lasers and blank cartridges to simulate actual battle.

Vehicles are typically outfitted with a belt of laser sensors or individual wireless detectors. Dismounted soldiers often wear a vest or harness with sensors as well as a "halo" of sensors on their helmets. MILES systems can be coupled with a real-time data link allowing position and event data to be transmitted back to a central site for data collection and display. More sophisticated systems for tanks and APCs exist that use various techniques (including scanning lasers and coupled radio systems) to allow more precise targeting of armored vehicles.

Various views of the M2A3 turret front and TOW launcher. The ABOVE photo shows the TOW launcher elevation mechanism, MILES sensors on the olive green strip attached with velcro across the front of the turret, and the smoke grenade launchers on the lower turret just above the hull.

ABOVE: A close-up view of the front of the TOW launcher in the lowered or travel position. The launcher is never raised while the vehicle is in motion. As the launcher is raised, the front cover is automatically opened. BELOW: The side of the TOW launcher that has the canvas cover on the back of the launcher.

Various views of the turret of the M2A3 Bradley. Note the additional armor on the turret above, with the rough anti-skid surface. These vehicles have a few bumps and paint scratches. CARC Tan paint is very durable.

ABOVE: A top view of the turret showing the Bradley's bustle rack on the rear of the turret. The bustle rack is relatively small, which is why many crews stow their rucksacks as well as many other items such as MREs on the side of the vehicle. The top of the ammo box containers can be seen on the outside of the bustle rack.

Several overhead views of the crew hatches and the top front part of the turret. Note the anti-skid texture on the gunner's hatche as well as the top of the turret.

PAGE 64 - 65: Various views of the rear of the Bradley, showing the inside of the bustle rack. These photos show how scuffed up the rear of the turret can get inside the bustle rack. A few of the bolts are missing that secure the additional armor on the turret. These photos show the antenna mount, the ammo can racks, and the gunner's and commander's hatches. The ammo box containers have holes to allow water drainage.

A close view of the rear of the TOW Launcher. The loader in the back loads the TOW missile from this end of the launcher and secures them in place with the black lever.

M2A3 3rd Infantry Division, 1-64 Armor, C12 “Army of Darkness”

M2A3 3rd Infantry Division, 1-64 Armor, C12
"Army of Darkness"

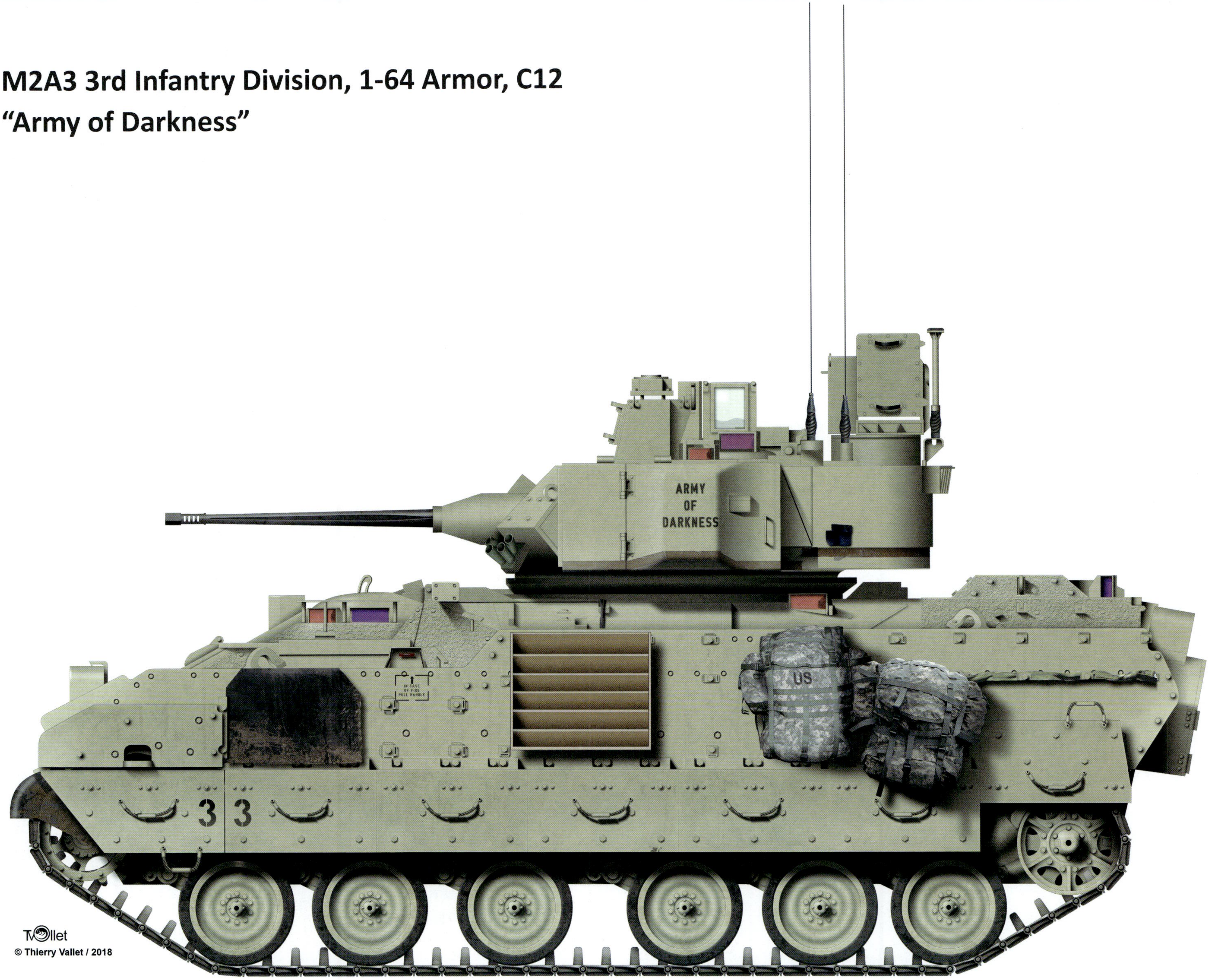

M2A3 3rd Infantry Division, 1st Brigade HQ66 "Prometheus"

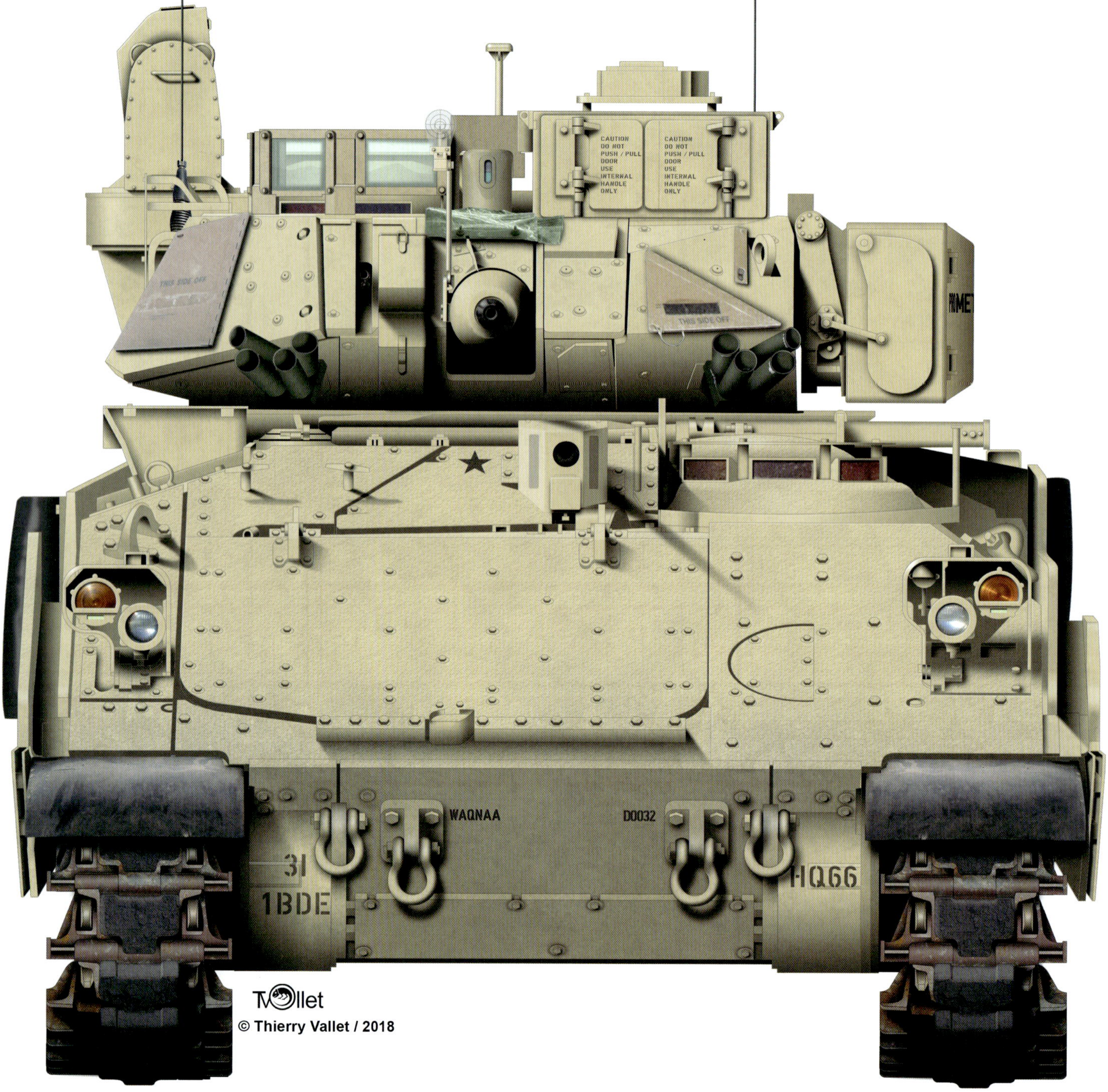

M2A3 3rd Infantry Division, 1st Brigade
HQ66 "Prometheus"

TVallet

Inside the Turret

The gunner's station of the M2A3

The commander's station of the M2A3

J3
J2
ENT
CIV
ON/UNSTOW
STOW
IBAS
AUTOTRACK
TRACK
DROP TRACK
POINT
GATE SIZE
DISPLAY
FOCUS
RETICLE
BRT
IMAGE BRT
IMAGE CON
ADJUST
LRF RETURN
FIRST
LAST
FLIR / TV ADJUST
MODE
FLIR
TV
POLARITY
WHT
BLK
LEVEL
GAIN
SCAN RATE
FILTER
ZOOM
ADJ MODE
MAN
AUTO
TAS ALIGN
FC
BACKUP ON
NORMAL
TURRET
POWER
ON
OFF
GRENADE
LAUNCHER
POWER
ON

The commander's station of the M2A3

CONNECTOR P2 MATES
CLOSED
OPEN
MADE IN U.S.A. MFR 95270

MANUAL
POWER
ELEVATION
DRIVE
Channel Lock Pliers
Screwdriver
Ratchet Wrench

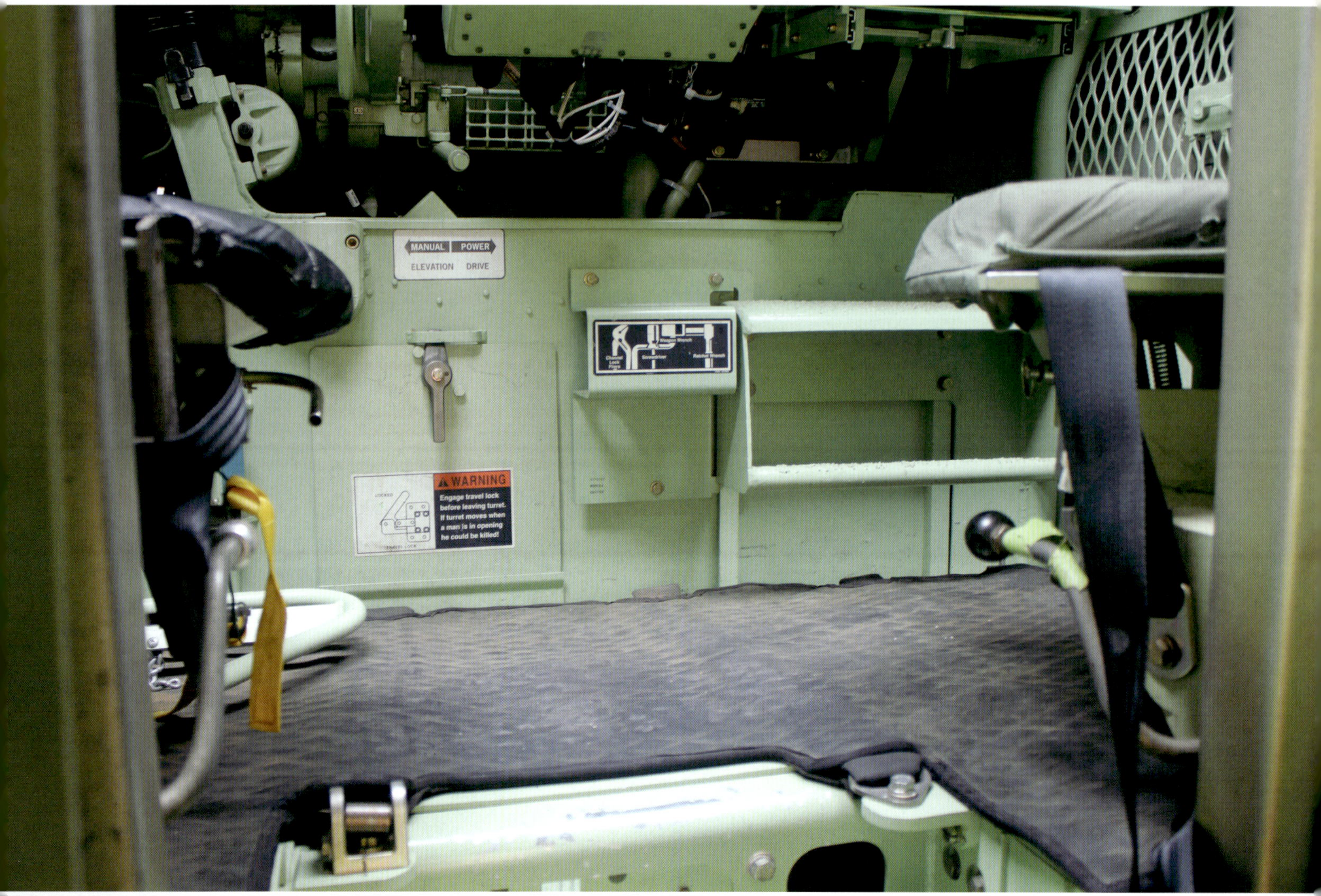
MANUAL
POWER
ELEVATION
DRIVE
Channel Lock Pliers
Screwdriver
Ratchet Wrench
WARNING
Engage travel lock before leaving turret. If turret moves when a man is in opening he could be killed!

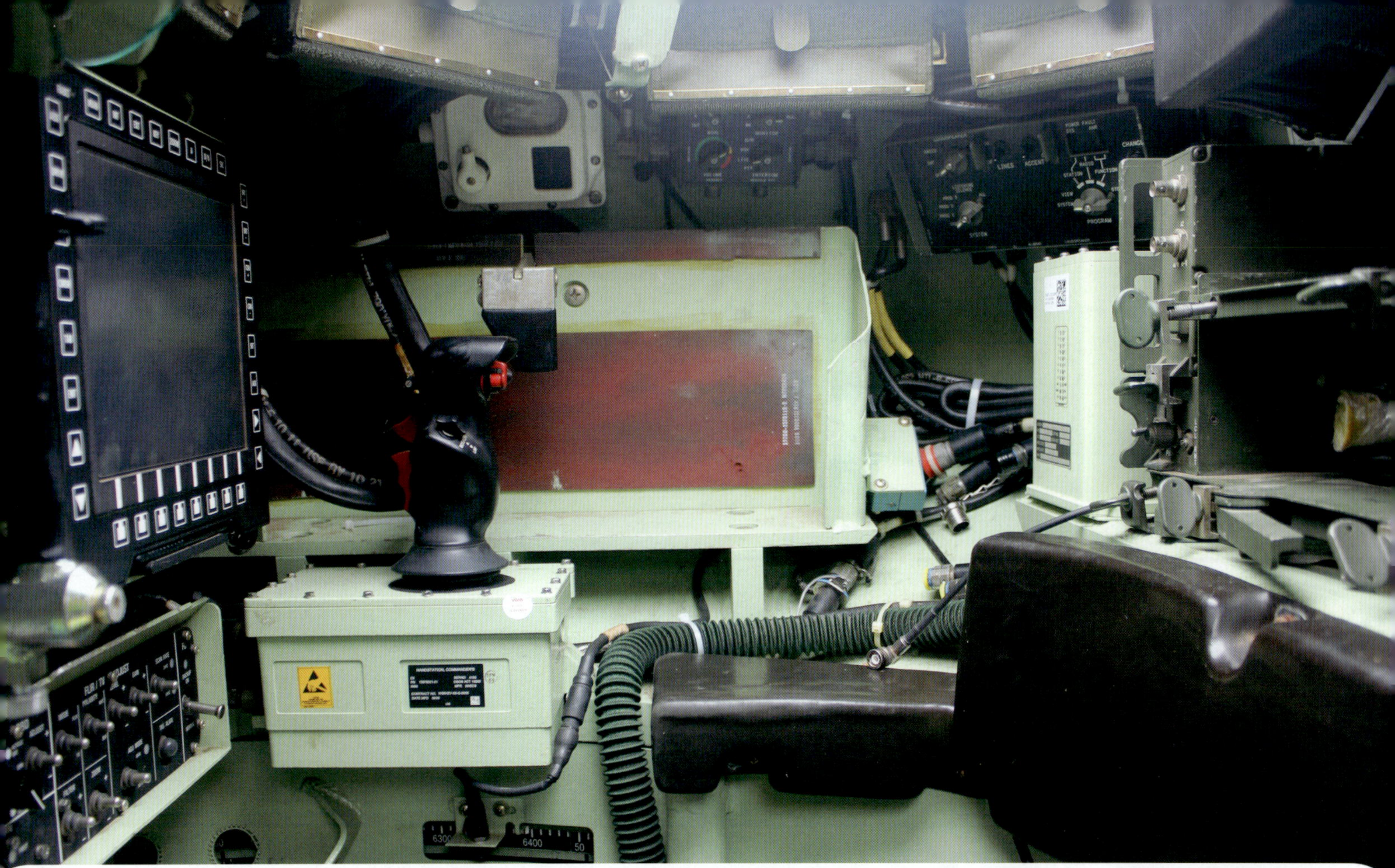

The commander's station of the M2A3

Looking down into the gunner's station.

ABOVE: Commanders override.
BELOW: 25mm chain gun ammo feed.

Inside the Rear

ABOVE: Crew compartment, left side.
BELOW: Crew compartment, right side.

ABOVE: Turret door, closed.
BELOW: Crew compartment, right side.

ABOVE: Crew compartment, left side.
BELOW: Crew seats with seat belts.

TO OPERATE
METALCRAFT
OPEN
CLEANING OR HOSING DOWN

Driver's Hole

ABOVE: Drivers controls.
BELOW: Driver's compartment.

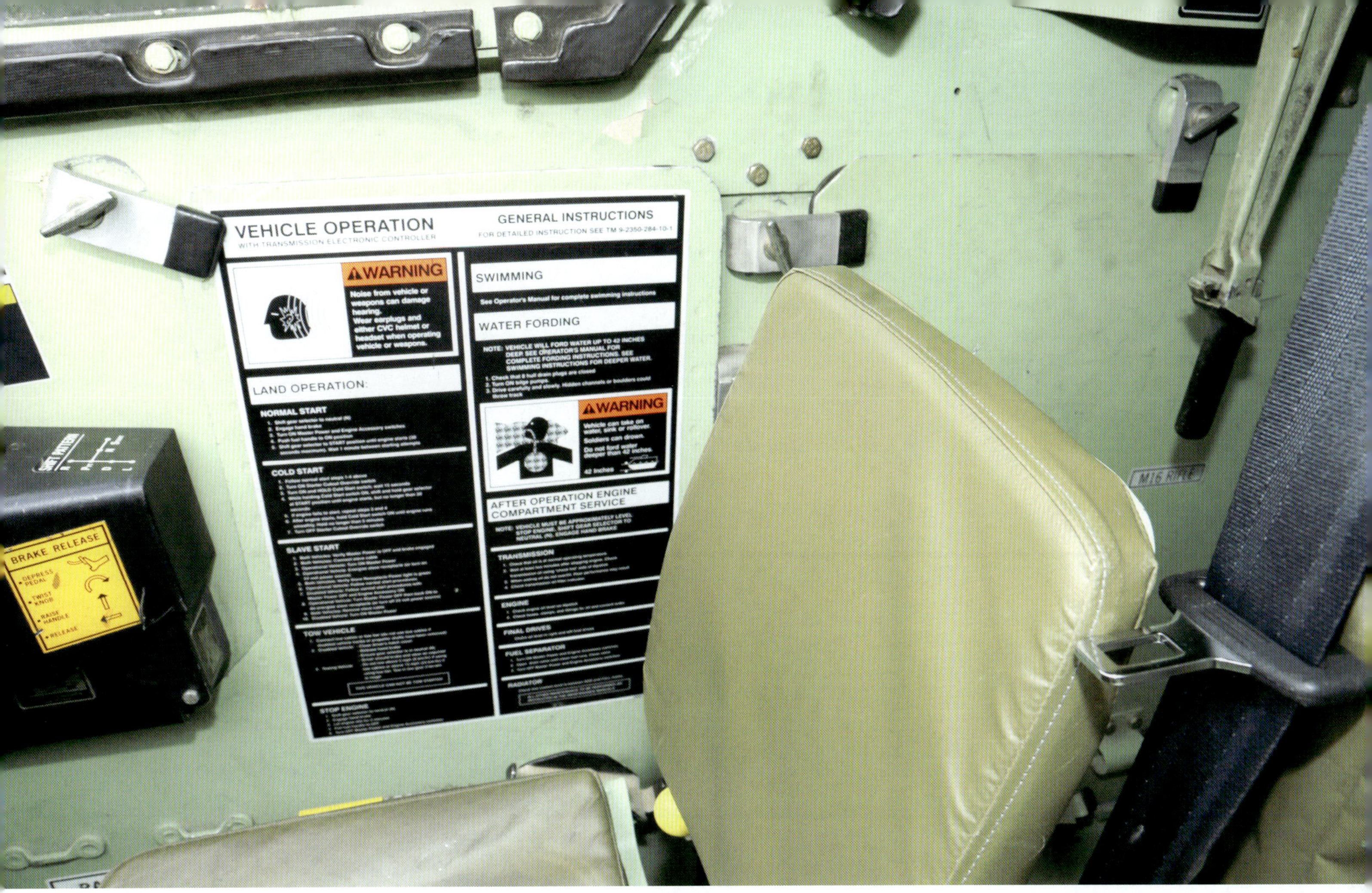
VEHICLE OPERATION
WITH TRANSMISSION ELECTRONIC CONTROLLER
GENERAL INSTRUCTIONS
WARNING
Noise from vehicle or weapons can damage hearing. Wear earplugs and either CVC helmet or headset when operating vehicle or weapons.
LAND OPERATION:
NORMAL START
COLD START
SLAVE START
TOW VEHICLE
STOP ENGINE
SWIMMING
See Operator's Manual for complete swimming instructions
WATER FORDING
WARNING
42 inches
AFTER OPERATION ENGINE COMPARTMENT SERVICE
TRANSMISSION
ENGINE
FINAL DRIVES
FUEL SEPARATOR
RADIATOR
BRAKE RELEASE
M16 RIFLE

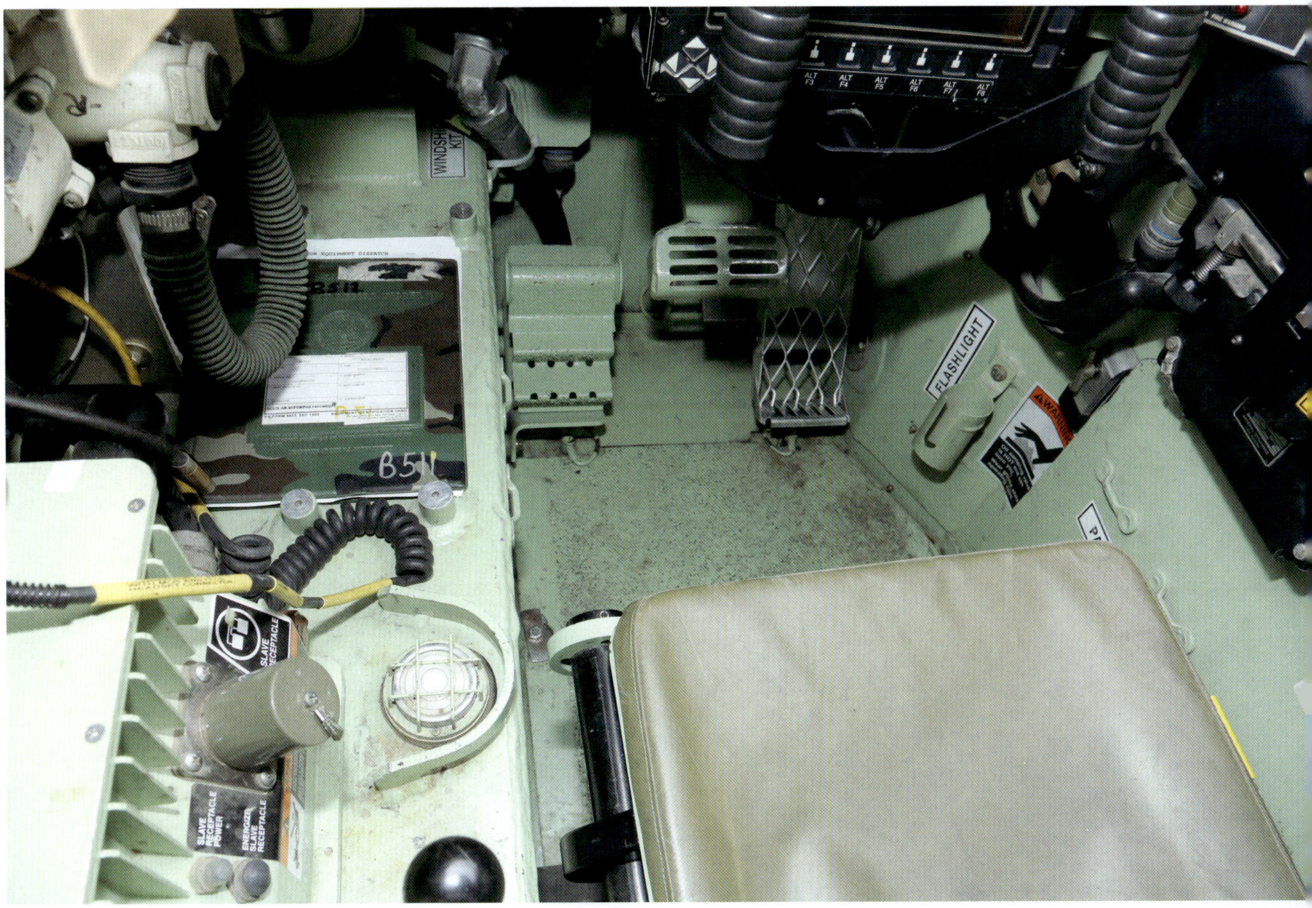
WINDSHIELD KIT
B511
FLASHLIGHT
SLAVE RECEPTACLE

The M2A3 Gunnery at Ft. Stewart

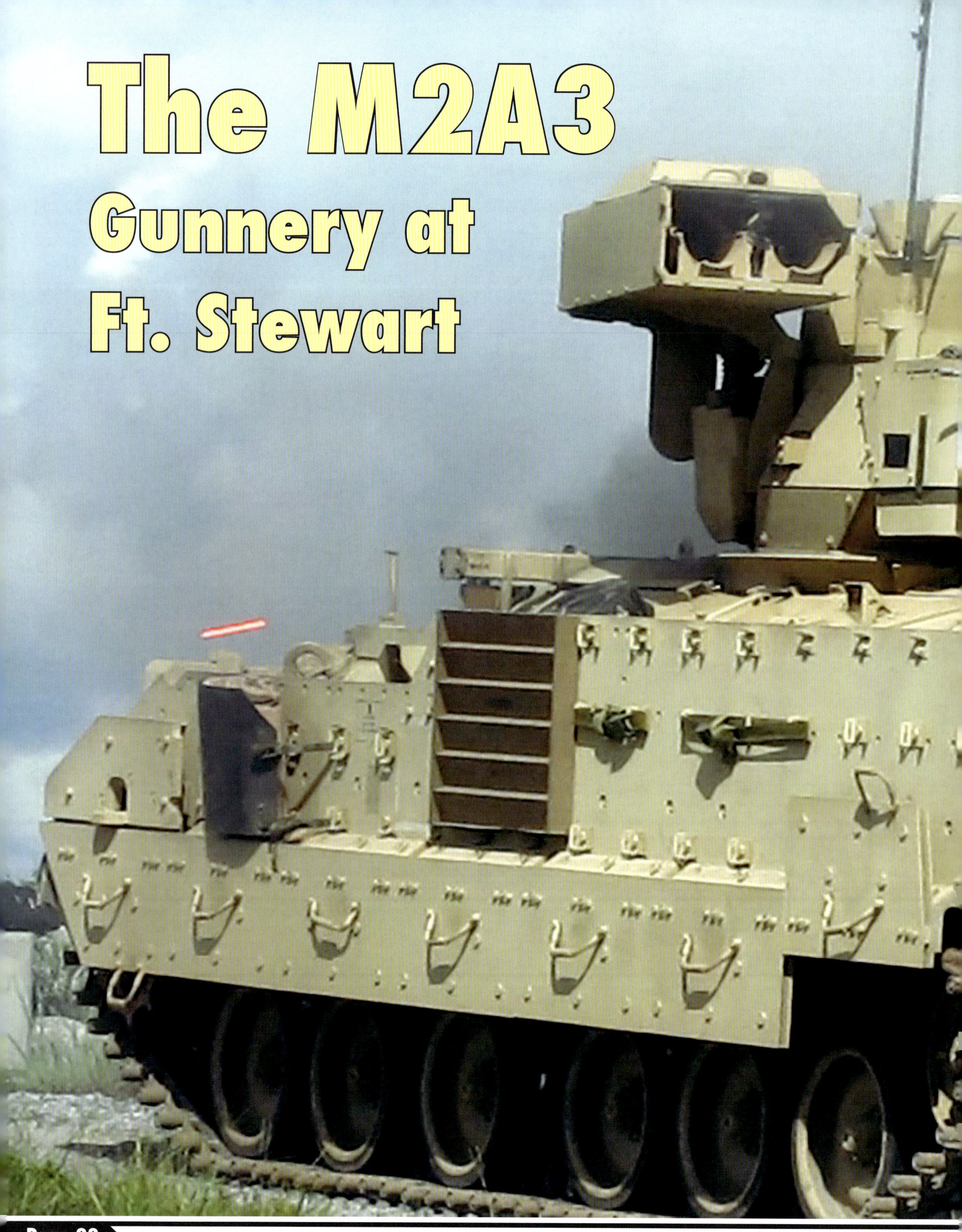

31
1-64
B21
WAZBBO
00005

ABOVE: Thc crew stands ready for the call from the bullpen. BELOW: The crew perform some basic preventive checks and maintenance on their vehicle in the motorpool.

ABOVE: This Bradley approaches BP1 to get ready to start the gunnery engagement. The main gun is in the fully elevated position until it is time to engage. BELOW: Ready to engage the target.

M2A3 Bradley
Concept Art
199th Infantry Brigade, Ft. Benning
OCP camouflage

M2A3 Bradley
Concept Art
199th Infantry Brigade, Ft. Benning
OCP camouflage, Large Pattern

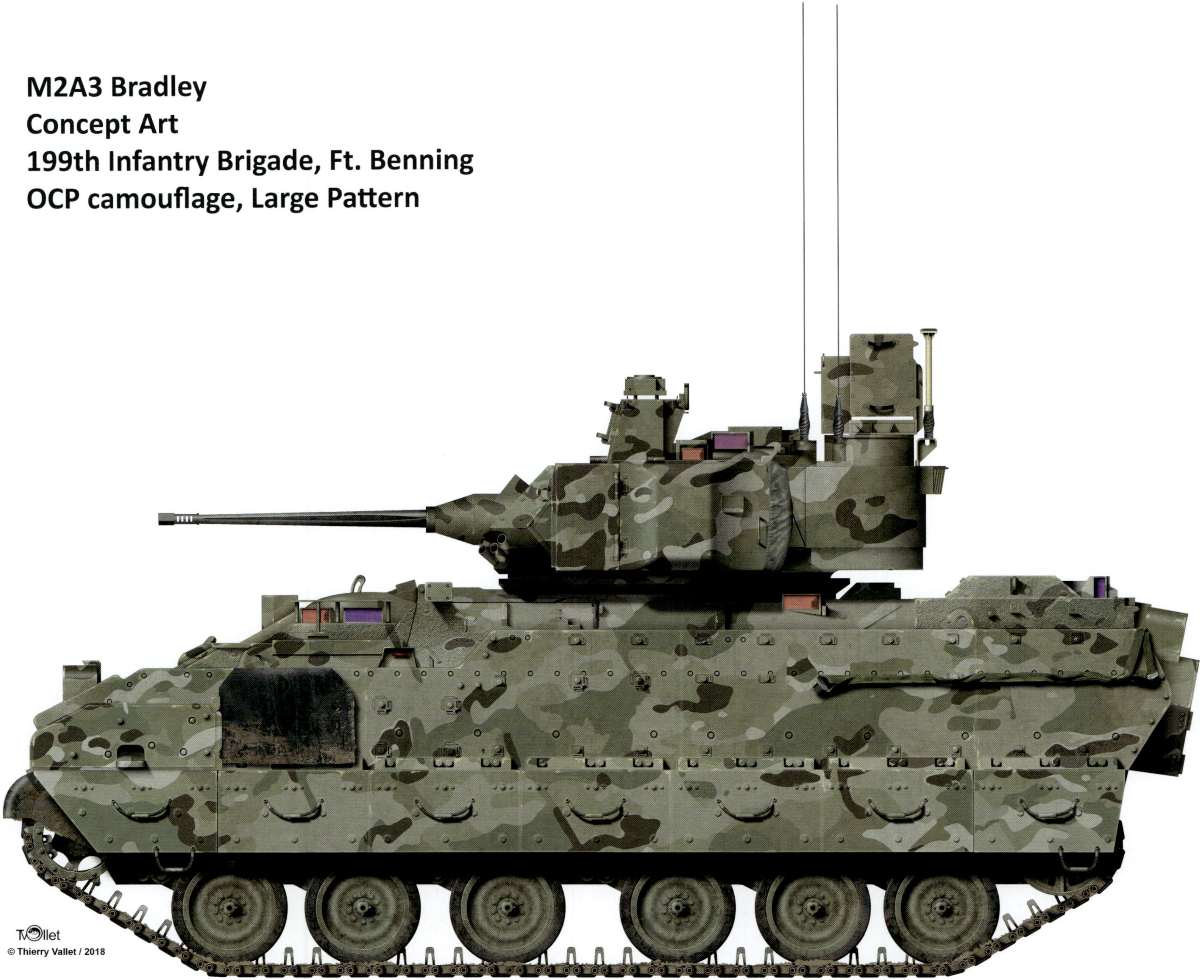

The M2A3 In the Field Ft. Stewart

C14
THIS SIDE OFF
3

THIS SIDE OFF
CONTEGNI

IN CASE
OF FIRE
PULL HANDLE
C14
TURRUBIATEZ

ABOVE: The ramp goes up while the dismounts provide security before bounding onto their attack position. Previously, the BC threw smoke to help disguise the dismounts movements.

BELOW: Now returning from assaulting the trench, one dismount provides security while another mounts up.

THIS SIDE OFF

33
31
-64

A BFIST Bradley returns from a successful attack. The bumper code indicates this Bradley Fire Support Team (B-FIST) vehicle is from 1-41 FA (Field Artillery).

C22 speeds into position for an attack during a CALFEX. The left front mudflap has been damaged.

The dismounts are fully loaded into the track after their assault and are waiting for the ramp to close up.

ABOVE: One of the last dismounts, the squad leader, enters the troop compartment. He has a dayglow flag to alert friendly forces of his squad's presence in an enemy position (the trench they recently assaulted).

BELOW: The dismounts return from their assault. During that time this M2A3 and others were providing overwatch for the assaulting troops.

Maintenance

Page 106-107: The M2A3 turret on its stand for maintenance.

AP AMMO LOADING (FULL LOAD 70 ROUNDS)
1. COUNT A STRING OF 70 ROUNDS
2. LOAD FIRST ROUND AS SHOWN.
3. ADD LOOPS AS SHOWN.
4. FEED OTHER END OF STRING THROUGH FORWARDER TO CANNON.
FIRST ROUND FOR LOADING
SINGLE EMPTY CLAW
FORWARDER
TURRET FRONT

ABOVE & BELOW: What's underneath the turret when it is removed from the Bradley. Note the grease and oil, and just plain dirt, tha accumulates from constant field maneuvers.

PAGE 109 ABOVE: The Bradley still has the tow bar attached. BELOW: The Bradley turre separated from the hull for maintenance. The photo shows the chain on the top of the turret that was used to lift the turret off the hull.

PLYMOVENT
USED OIL

The powerpack of the M2A3 removed from the hull ready for maintenance.

The right side of the engine compartment.

The left side of the engine compartment.

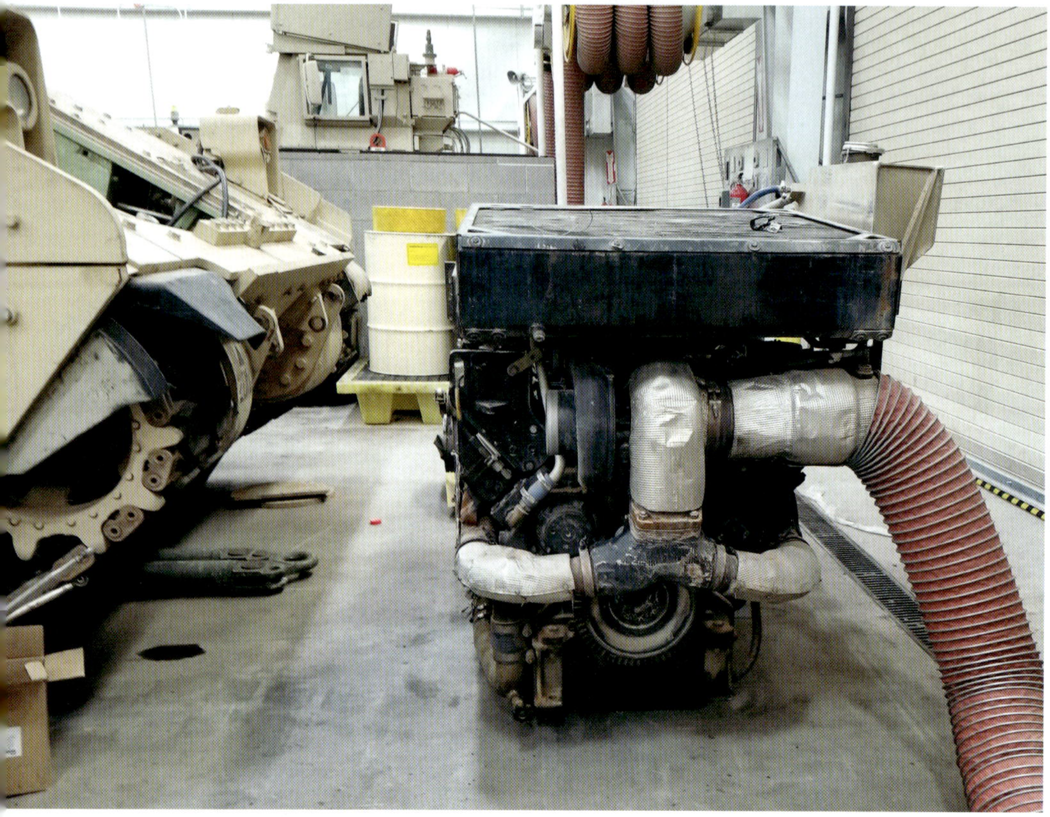

The powerpack on the Bradley consists of a Cummins VTA-903T 8-cylinder diesel producing 600 hp (447 kW). The Bradley can reach speeds of over 40 miles per hour and has a range of over 300 miles. The Bradley possesses sufficient cross-country mobility to keep up with the M1A2 Abrams Main Battle Tank.

The M242 Bushmaster chain gun.

BCOTM Bradley

Battle Command on the Move (BCOTM) is a revolutionary capability that provides current and future combined arms commanders all the information resident in their command posts, and the required communications necessary to command and control their combined arms team on the move, or at a short halt, from any vantage point on the joint battlefield.

062

WANMAA0084
41D 2 BCT

WANMAA0084

M2A3 Bradley at NTC 2008

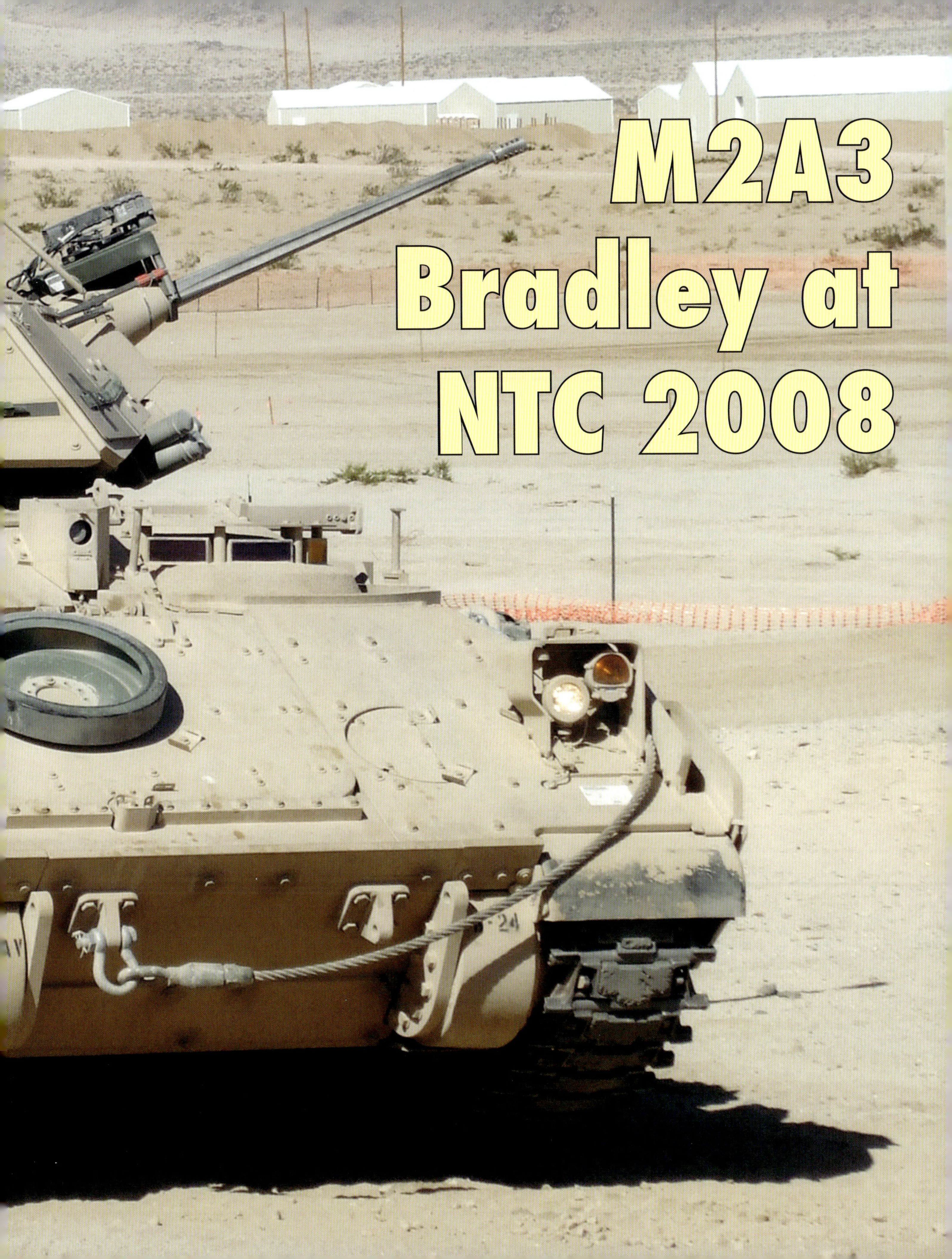

1CAV 2-5CAV
B-14

1CAV
2-5CAV
B-50

M2A3 Bradley
10th Engineers, 3rd Infantry Division
Vehicle B12
Head Hunter

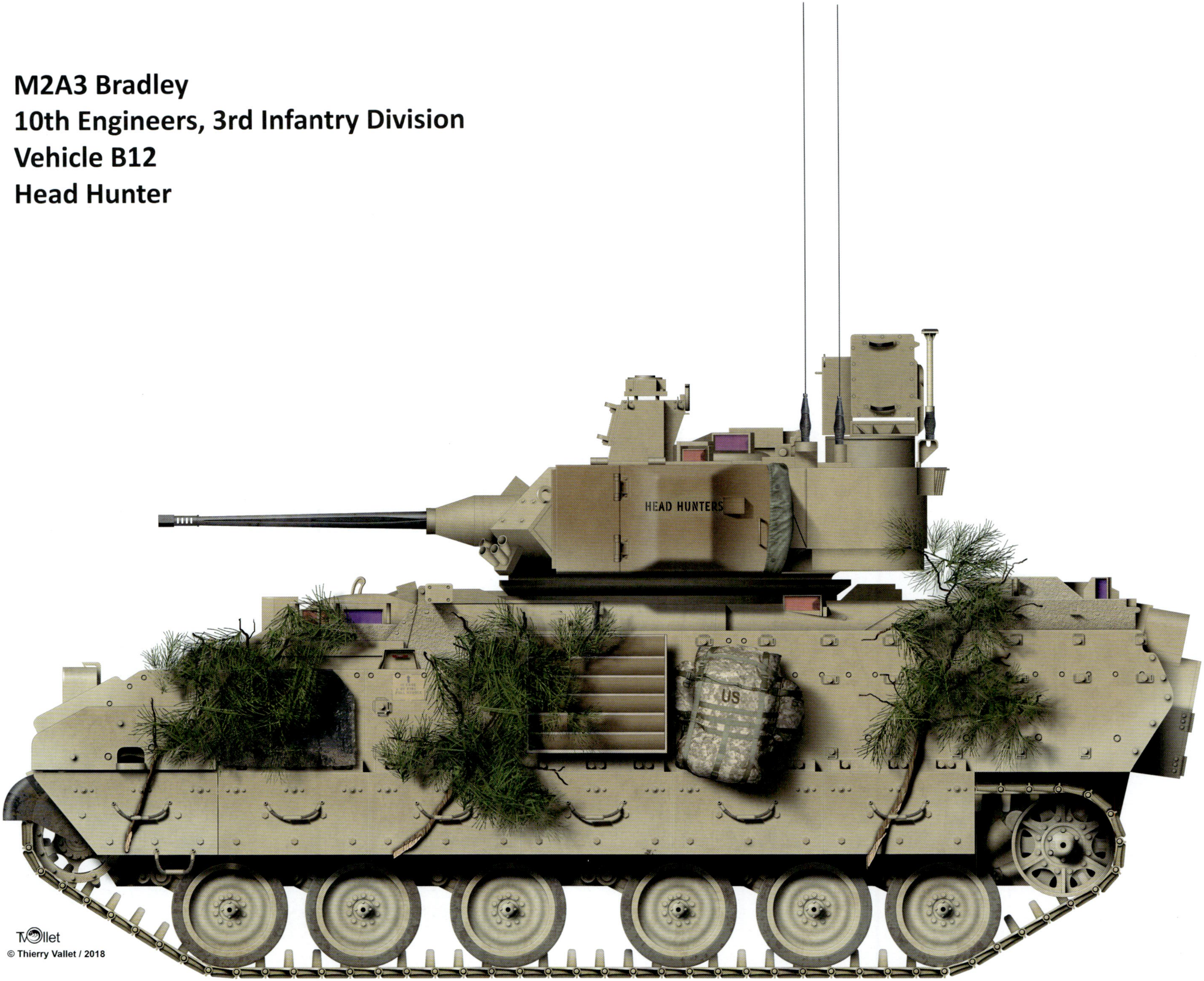

M2A3 Bradley
10th Engineers, 3rd Infantry Division
Vehicle B16
Bane

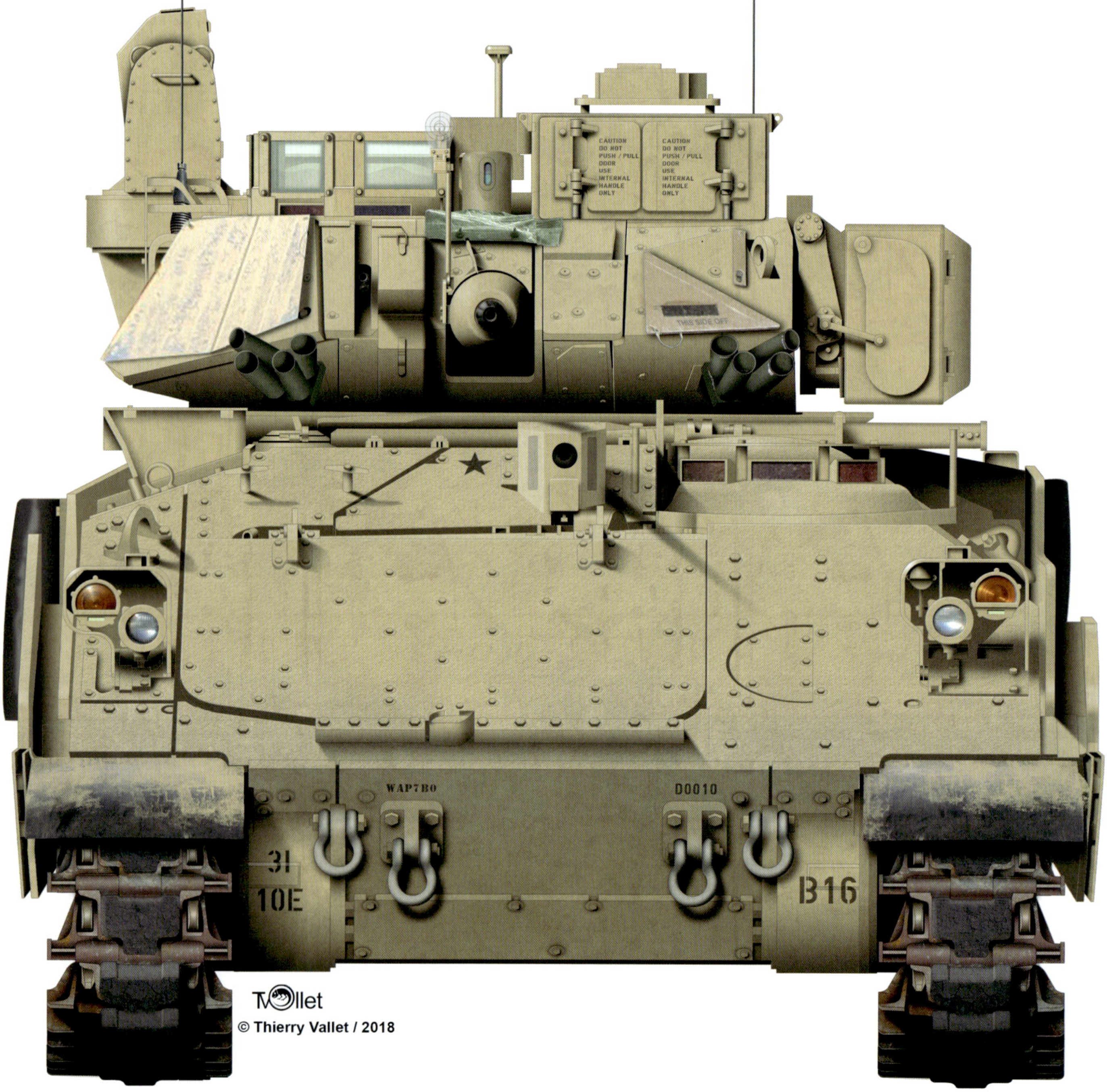

M2A3 Bradley
10th Engineers, 3rd Infantry Division
Vehicle B16
Bane

Boneyard Bradley

M2A3 Bradley
1-5 Cavalry, 1st Cavalry Division, Korea
October 2017

Bradleys in Korea

RODRIGUEZ LIVE FIRE COMPLEX, Republic of Korea: Bradley Fighting Vehicles from Alpha Company, 8th Brigade Engineer Battalion return after conducting breaching operations with 1-9CAV on August 10, 2017.

An Alpha Company, 8th Engineer Battalion Bradley moves along a road while training at the Korean Training Area/Rodriguez Range Complex in South Korea.

A fully laden M2A3 Bradley from 1-9 Cavalry, 1st Cavalry Division, attached to the 2d Infantry Division, moves through a Korean village. Note the amount of baggage that is strapped to the outside of the vehicle.

M7A3 B-FIST at Ft. Sill 2016

The Bradley Fire Support Team Vehicle or M7A3 B-FIST is outwardly similar to its cousin the M2A3 Bradley Fighting Vehicle. Built from the M3A3 Cavalry Fighting Vehicle, it has side and rearward facings vision blocks integrated into the TOW Loading Hatch instead of between the hatch and the ramp. Additionally, the TOW Missile Launcher is replaced with a larger armored box which mounts the Fire Support Sensor System (FS3) that allows the crew to identify and laser designate targets to be engaged with artillery (conventional or precision guided rounds).

PAGES 138 - 139: These 4 images show off the main external difference between the M2A3 Bradley and the M7A3 Bradley Fire Support Team Vehicle: The FS3 Fire Support Sensor System. This box has a slightly different shape than the TOW2 missile launcher and is a little larger. You can also see the access panel open on the FS3 and see the cables to the laser rangefinder.

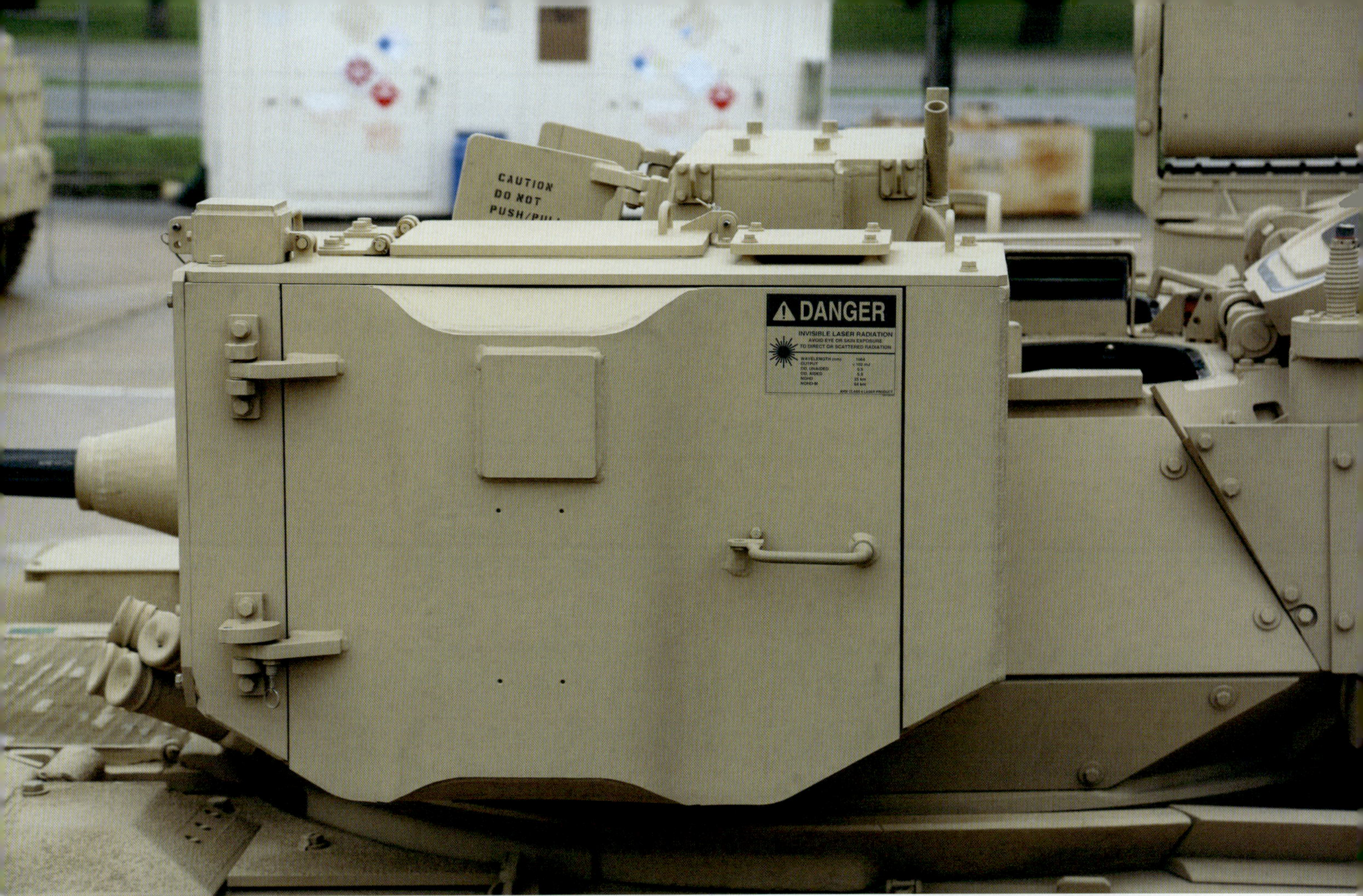
CAUTION
DO NOT
DANGER
INVISIBLE LASER RADIATION

CAUTION
DO NOT

PAGES 138-139: The access panel on the FS3 can be seen in the open position along with the cable to the laser rangefinder.

Infrared Laser Radiation Emitted From This Aperture
DANGER

DANGER
INVISIBLE LASER RADIATION

The crew compartment of the B-FIST is very similar to the M2A3, except for the targeting and communications equipment on the right side.

FIRE
EXTINGUISHER

FIRE
EXTINGUISHER
GOGGLES

ABOVE & LEFT: Engineers with the 116th Brigade Engineer Battalion (BEB) conduct M2A3 Bradley fighting vehicle gunnery qualification on March 27, 2018, Orchard Combat Training Center, south of Boise, Idaho. Combat engineers with the 116th BEB trained through gunnery table XII, evaluating their ability to execute collective platoon-level tasks in a tactical live-fire environment; including integrating dismounted soldiers with their assigned BFV.

U.S. Army Soldiers assigned to 1st Brigade Combat Team, 1st Cavalry Division, defend their position in an M2A3 during Decisive Action Rotation 18-02 at the National Training Center in Fort Irwin, Calif., November 22, 2017.

M2A3 Bradleys belonging to the 1st Battalion, 155th Infantry Regiment, are staged for refueling May 22, 2017 at the National Training Center, Fort Irwin, California.

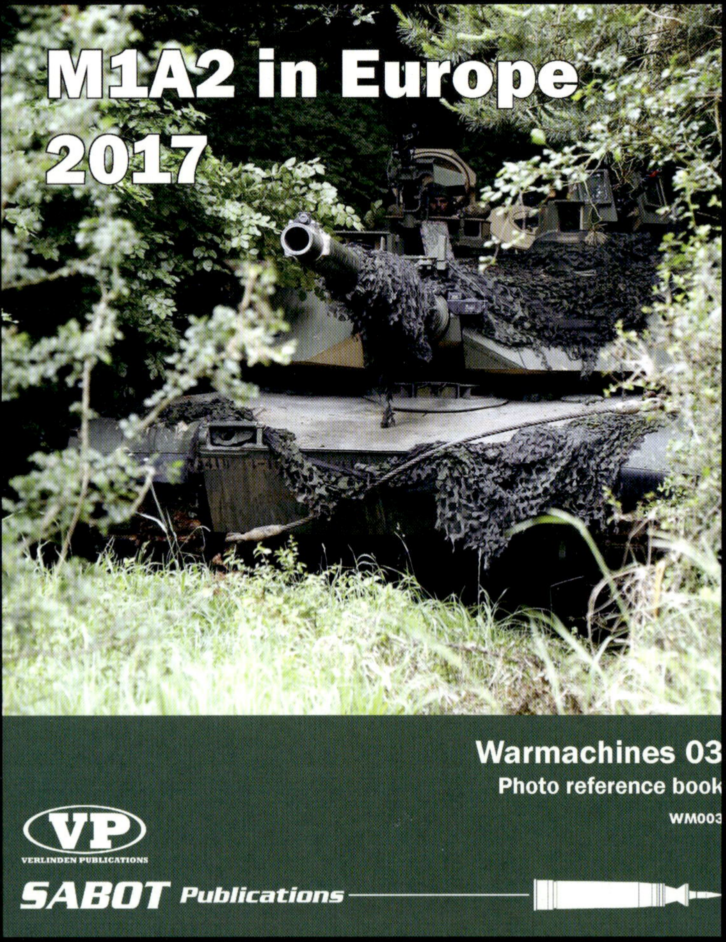
M1A2 in Europe
2017
Warmachines 03
Photo reference book
WM003
VP
VERLINDEN PUBLICATIONS
SABOT Publications

M1A2 SEP
Abrams Main Battle Tank
In Detail
The Revised and Expanded edition of SABOT Publications first book on the M1A2 SEP. Now with more than 140 pages, detailing in photographs the U.S. Army's Main Battle Tank.
SABOT Publications
SP007

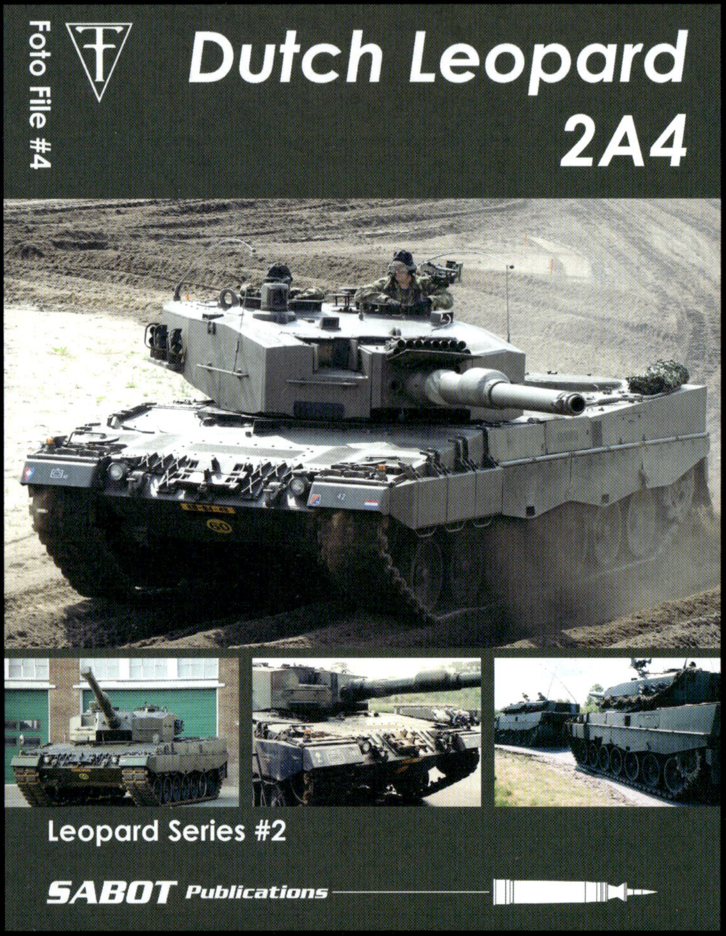
Foto File #4
Dutch Leopard
2A4
Leopard Series #2
SABOT Publications

M1A1 SA in Iraqi Service
Includes Bonus
Decal Sheet!
Warmachines 04
Photo reference book
WM004
VP
VERLINDEN PUBLICATIONS
SABOT Publications